Theory of Structure In AI 4

By

Ian Beardsley

© 2018 by Ian Beardsley

ISBN: 978-1-387-63409-5

Table of Contents

AI And The Biological

Ian Beardsley

It would seem that automaton, or computers, or
artificial intelligence (AI) was developed by imitating
the human central nervous system. Nerve cells send a
pulse from the brain to say, a muscle telling it to contract, so
the user can tell himself to say, move his arm. In the same way
a user presses a key on the keyboard sending a message to the say,
processor to say, type the letter pressed on the screen. As such,
it has been shown that electric neural networks of aggregates of black boxes
can be put together such that they can represent any logical principle.
It is my belief the first thing we should do in AI, if we want to build
from the ground up, is find the connection between the chemical
composition of artificial intelligence to the chemical composition of
biological life. I have done this. But, it would seem, computers got their
start by the structure of their circuitry after the human central nervous system.
My books, Theory of Structure in AI, Theory of Structure in AI 2, and
Theory of Structure in AI 3, outline my work in the former and, I am
currently working on Theory of Structure in AI 4 which examines the
latter.

The Ancient Greeks believed proportions were the ultimate expression of nature. It means that;

a/b=c/d

a is to b as c is to d

What I found is that such a proportion exists in transistor technology, the basis of artificial intelligence (AI) and, the biological through the golden ratio and means geometric, arithmetic and harmonic.

I have found that there is structure in artificial intelligence by the main materials used to make transistor technology Silicon (Si), Phosphorus (P), and Boron (B) through the golden ratio called phi ($\phi=0.618$) as follows:

$$\frac{\sqrt{PB}\,(P+B)+2PB}{2(P+B)Si} \approx \phi$$

And, that that structure is in biological life which is chiefly composed of Carbon (C), Nitrogen (N), Oxygen (O), Hydrogen (H) as follows:

$$\frac{C+N+O+H}{P+B+Si} \approx \phi$$ Which means AI is connected to the biological through the golden ratio as such:

$$P+B+Si\,\frac{\sqrt{PB}\,(P+B)+2PB}{2(P+B)Si} \approx C+N+O+H$$

This is the equivalence between AI on the left side of the equation and, the biological which is on the right side of the equation.

Ultimately, I have discovered the following connecting AI to DNA, Amino Acids, mineral component of bone, and on...

Equations Set 1

AI Form	Biological Component	Biological Function
$\dfrac{\sqrt{PB}\,(P+B)+2PB}{2(P+B)Si}$	$\dfrac{H_2O}{Air}$	primary life sustenance
$\dfrac{1}{Si}\left[\sqrt{PB}+\dfrac{2PB}{P+B}+\dfrac{P+B}{2}\right]$	$\dfrac{H_2O}{Air}\left[\dfrac{H_2O}{CH_4}+\dfrac{NH_3}{CH_4}+1\right]$	precursors to amino acids
$\left[\dfrac{P}{Si}+\dfrac{B}{Si}+1\right]$	$\left[\dfrac{2CH_2+CO+NH_3}{CH_2+OH}\right]$	amino acids (build proteins)
$\left[\dfrac{Ga}{Ge}+\dfrac{As}{Ge}+1\right]$	$\left[\dfrac{C_5H_5N_5}{C_3H_6N_2O_2}+\dfrac{C_5H_5N_5O}{C_3H_6N_2O_2}+\dfrac{C_4H_5N_3O}{C_3H_6N_2O_2}\right]$	organic bases (DNA)
$\left[\dfrac{Se}{Zn}\right]$	$\left[\dfrac{C_5H_6N_2O_2}{H_3PO_4}\right]$	phosphoric acid (DNA)
$\left[\dfrac{Ge}{Si}\right]$	$\left[\dfrac{C_{17}H_{91}N_{19}O_{16}}{Ca_5(PO_4)_3(OH)}\right]$	organic component bone / mineral component bone
$-2cos\left(\dfrac{\pi}{4}\right)+2cos\left(\dfrac{\pi}{5}\right)+2cos\left(\dfrac{\pi}{6}\right)\dfrac{Zn}{Se}$	$\dfrac{air}{H_2O}$	primary life sustenance

Early Computer Science

The General And Logical Theory of Automaton, by John Von Neumann

The reason I studied this, is that it is about how automata computing machines function as the biological neural network and my theory (Theory of Structure in AI, Theory of Structure in AI 2, and Theory of Structure in AI 3) shows inherent structure in the chemical composition of artificial intelligence (AI), and that that structure is corollary to the chemical composition of biological life.

To understand the science of artificial intelligence (AI) we look at the vocabulary initiated in the early years of its development by referring to an essay published in the four volume set "The World of Mathematics" published in 1956 by Simon and Schuster,...

. Automaton
. Computing Automaton
. Analogy Machines
. Digital Machines
. Dynamometer
. Rheostat
. Signal to Noise Ratio
. Decimal
. Binary

Automaton

    Computing Machines

        Analogy Machines

        Digital Machines

Analogy Machines

    Operations represented by a current or the rotation of a disc through an arc (something mechanical).

    Multiplication of two currents

    Feed currents into the two magnets of a dynamometer producing a rotation of a disc and transform that into an electrical resistance by attaching a rheostat. The resistance can be converted into a current by putting a potential across it.

        The aggregate is a "black box" that produces a current proportional to the product of the two inputs.

    Signal to noise no lower than 1/100,000
    As high as 1/100

The essay moves on to draw a comparison between the biological organisms and digital machines. It uses what we call today "lumped element abstraction" but rather refers to it as "aggregates" of "black boxes". Lumped element abstraction is quite simply, instead of solving Maxwell's Equation for a light bulb, say…

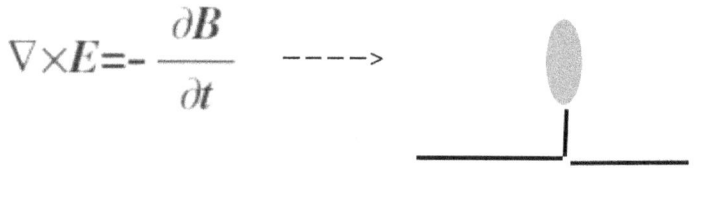

We write the schematic,…

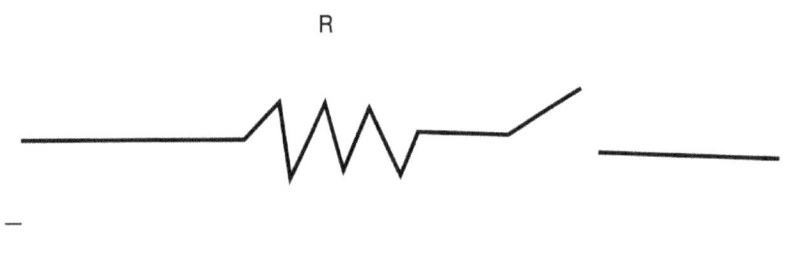

$$V = IR$$

In other words, since a neuron is a biological organ that can be considered a switching organ or relay organ, the same as a vacuum tube is a switch or relay, it can be taken as such regardless of the fact that both are really analogy machines.

Von Neumann goes on to correctly propose that digital machines will show more promise that analogy machines, and he comes to this conclusion pointing out that the signal to noise ratio is much higher in the digital computing machines than in the analogy computing machines. What we mean by a digital machine is that any number can be represented by two or more numbers, such as binary, which is base 2, a zero or one, an on or a off. That is if a switch is off, it represents 0 volts, if it is on it represents 1 volt. This said, he says that we can consider that the neuron sending an impulse in the biological organism can be taken as a digital organ if considered as a discrete whole element, even though its parts function independently as analogy machines.

The essay was written at the time when we were building the first computing machines, the ENIAC in Aberdeen, Maryland and, the SSEC in New York owned by IBM. Both of which consisted of about 20,000 switches which were vacuum tubes for the ENIAC and vacuum tubes and electromechanical relays for the SSEC.

He points out that while these machines were enormous compared to the biological neural network because, a vacuum tube is enormous compared to the size of a nerve cell (about a billion times larger) the digital machines were much faster with 1,000,000 actuations per second and, the nerve cell with 1/200 of a second for stimulation to recovery, which was a ratio of 1:5000.

Thus, today when we use binary, each character on the key of a computer keyboard is represented by a binary number which is a byte, or eight bits, where a bit is a zero or a one. Thus,...

Binary
0 = 0   1 = 1   10 = 2   11 = 3   100 = 4   101 = 5   110 = 6   111 = 7,...
97 = 11000001 = a   98 = 1100010 = b   99 = 1100011 = c,...

And, if we want to print something, instead of typing something that looks like:

We develop a computer language where printf() represents that in a library. Thus, in C...

print.c

#include (stdio.h);	define the library
int main (void);	code begins here
{	code to follow
printf ("hello")	

Compile and run with linux command line interface: make print, ./ print

hello

Thus, since we now have transistor technology to make a switch,…

…in place of vacuum tube technology

This,…

Becomes,…

Let us state more clearly what analogy and digital machines are. Analogy machines are mechanical, such as the measure of the arc of the rotation of a disc representing the product of two numbers, and a digital machine uses switches that are either on or off, that is, is something like a switch either being on or off, a yes or a no. That is, we say a nerve cell in a biological organism is like a digital organ because it sends an impulse from the brain, no matter its size, to either contract a muscle in the body or not and, a vacuum tube is digital because it is a switch that is either on or off.

But a nerve cell as a digital organ is only an abstraction that can be taken if we consider the end result of its action as a discrete element. Really it is an analog machine because we must consider the source of the impulse, which is the energy released by a complex chain of biochemical reactions. Similarly, the vacuum tube as a digital organ is an abstraction, as well, because we must consider the source of it being either on or off, which is a potential across its cathode and anode supplied by an electromotive potential, such as a battery.

Since my theory is that there are parallels between the chemical structure of the materials that make AI and the chemical materials that make biological life, we are interested in what these materials really are, not their abstractions. That is, we are interested, at the present, in analogy machines, to build the foundation of the theory of the connection of AI to the biological. After that, we can consider the relationship between AI and the biological, in their digital behaviors, as the next step in formulating the theory.

Electron Volt: A unit of energy equal to the work done on an electron in accelerating it through a potential of one volt. It is 1.6E10-19 Joules (Google Search Engine)

Volt: Potential energy that will impart on joule of energy per coulomb of charge that passes through it. (Wikipedia)

Coulomb: The charge of 6.242E18 protons or 6.242E18 electrons.

Forward Bias: A diode (silicon) must have 0.7 volts across it to turn it on, 0.3 volts (Germanium). This is called forward voltage. The forward voltage threshold is 0.6 volts.

(0.6 volts)(1.6E-19)=9.6E-20 Joules

This is the energy to turn on a diode, or the threshold of life for artificial intelligence. I call it a bue (basic unit of energy).

```
+1.2eV — — — — —|>— — — — — — — — —> out (9.6E-20 J, or 0.6 eV)
                |
                |
      — — — — —|>— — —|
                |
                |
                R
                |
                |
      — — — — — — — — —
```

A photon has to have a minimum energy of 1.2 electron volts to impart to an electron for it to turn on the simplest of logic gates; a one on, one off, OR GATE, for there to be an output of one "bue" (basic unit of energy electronic) , which 9.6E-20 Joules, as I have calculated it.

Use Planck's Equation: E=hv where h= 6.626E-34 Joule seconds

v=2(9.6E-20)/(6.626E-34)=3.067E14 cycles per second

wavelength = lambda = c/v where c is the speed of light equal to 3E8 m/s

lambda = (3E8)/(3.067E-14) = 9.78E-7 meters

1 micrometer = 1E-6 meters

lambda ~ 1 micrometer (This is where the visible spectrum begins)

So we see the visible spectrum for one photon of light begins where the energy is 2 bue.

I further wrote the following in the same book,…

Aerobic respiration requires oxygen (O₂) in order to generate ATP. Although carbohydrates, fats, and proteins are consumed as reactants, it is the preferred method of pyruvate breakdown in glycolysis and requires that pyruvate enter the mitochondria in order to be fully oxidized by the Krebs cycle. The products of this process are carbon dioxide and water, but the energy transferred is used to break strong bonds in ADP as the third phosphate group is added to form ATP (adenosine triphosphate), by substrate-level phosphorylation, NADH and FADH2 (Wikipedia)

Simplified reaction:

$C_6H_{12}O_6$ (s) + 6 O_2 (g) → 6 CO_2 (g) + 6 H_2O (l) + heat
ΔG = −2880 kJ per mol of $C_6H_{12}O_6$

(Wikipedia)

This is the heat (enthalpy) of the combustion of C6H12O6

(2,880,000 J)/(6.02E23 C6H12O6) =4.784E-18 J = basic unit of biological life
(4.784E-18 J)/(9.6E-20 J)=49.8~50

This says the basic energy unit of organic, or biological life, is about 50 times greater than the basic energy unit of electronic life, or artificial intelligence.

That is 0.6(50)=30 electron volts = basic unit of energy for biological life.

So, we see the visible spectrum for one photon of light begins where the energy of the photon is 2 "bue" electronic which is 100 "bue" biological and that that photon has a wavelength of 1.0 micrometers.

This is all about vision in a robot or AI.

I now see we should do the same calculation we did for silicon diodes, for germanium diodes, which turn on at 0.3 volts. We have,…

(0.3 volts)(1.6E-19 Joules/volt)=4.8E-20 Joules

Thus, we see that twice the basic fundamental unit of energy to turn on the OR gate with an output of one basic unit of energy used to construct vision in a computer or robot (AI) is the energy of a photon of light that begins where the visible spectrum begins for humans. This visible spectrum for humans was acquired through millions of years of biological evolution.

Neural Pulses Encode By Way Of Frequency Modulation

If the nerve cell is a digital organ, that is sending pules no matter their size meaning only on or off, then how does it encode for strength? Von Neumann explains that it does so by frequency modulation. The time between pulses determines a frequency proportional to the desired strength the signals want to communicate. There are two possible approaches — either by binary, decimal, hexadecimal, …etc — expansion, or , by counting. While counting requires a million pulses to describe a million, expansion can do it in seven in decimal (there are seven digits in a million). He asks, why does nature choose the less economical notation of counting and, answers, because if there is an error in one of the bits the signal is still accurate enough to be viable whereas, if one bit is off in expansion, the entire signal fails.

Neural Networks

Von Neumann creates the following model for neural networks:

We view neuron as a "black-box" with a certain number of inputs that can be either excitatory, or, inhibitory, and that has an output. There are two kinds of boxes — Threshold 1, and Threshold 2. For an organ to be stimulated it must have for any instant as many stimuli on its excitatory inputs that correspond to its threshold, and no stimuli on its inhibitory inputs. If this is the case, it will, after a definite duration of delay, have an output pulse. This output can be connected to its own input, or, to the inputs of other "black-boxes". These can be connected to one another to construct formal neural networks. He writes of it:

"There is a connection between logical principles and their embodiment in a neural network."

This pertains to The McCulloch and Pitts Theory of Formal Neural Networks. That is, anything represented with words, art, or any other medium can be represented with a neural network.

Definition of a Computing Automaton

Now the essay gets at the crux of computer science. This is in the definition of a computing machine by the father of AI, Alan Turing. It is as follows,…

If an automaton can read a description of something and imitate what is described, then it is Universal. For example, if the principle behind a right triangle is encoded in binary on paper tape and fed into a computer, and that computer can read it, and then produce an image of it on the screen, or monitor then it is Universal. In other words, if,…

```
i=0;
while (i!=0):
      i=i+1;
      print (i times "#");
      print (" ");
```

run program,…

```
#
##
###
####
```

Preliminary Notes

Schematic for making a neural Network

sum over weighted
inputs

generate a weight random between 0 and 1 (a probability) $\longrightarrow$ $\sum_{i} x_i w_i$ $\rightarrow$ $||\sigma(x)||$ normalized activation formula $\rightarrow$ Compare with training set

$\rightarrow$ $\sigma(x)(1-\sigma(x))$ calculate errors (distance) $\rightarrow$ Present new problem

$\rightarrow$ Process with the learned weights

It is iterating a large number of times over the sum of the weighted inputs as generated by a random probability at $i=1$ that the system eventually learns the weights and biases for a problem ($i =$ large number) ofcourse the iterations over the sum have to be normalized

Neural Network

generate
random int (between 0 and 1:

$w = 2 * random.random((3,1)) - 1$
$\underline{\text{sum over the weighted inputs:}}$
$\sum x_i w_i = x_1 w_1 + x_2 w_2 + \ldots$

run through normalized activation

Formula:

$$\dfrac{1}{1 + e^{-\sum_i x_i w_i}}$$

Calculate distance between
computed values and actual:

~~error~~

$(error) \, \sigma'(x)(1 - \sigma(x)))$ ~~(scribbled out)~~

$error = (computed - actual)$

$$\sigma(x) = \dfrac{1}{1 + e^{-x}}$$

output = 1 /(1 + exp (-dot (inputs, weights))))
and using Python's neural network library
~~(scribbled out line)~~
~~(scribbled)~~ 1 /1 + np.exp (-model) = output

To generate a random number
in python:

```
import random
for x in range(1):
    print random.randint(1,11)
```

the first line imports the
random function
the second line tells how many
random numbers to generate
(in this case 1)
the third line tells between
what numbers should it generate
(in this case they ar 1-10)

to generate random numbers
between -1 and 1 for a
3 x1 matrix

```
randoms = 2*random.random((3,1)) -
```

You can seed the random
number generator (tell it what
integer to begin with in its
algorithm to produce a random
number)

```
        random.seed(1)
```
to produce same numbers each time

The sigmoid function

$$y = \sigma(x)$$

for large negative numbers it produces 0 and for large positive numbers it produces 1 (converges on 0 to left, one to right) Therefore it is perfect for normalizing the sum.

$$\sum_i x_i \omega_i = x_1 \omega_1 + x_2 \omega_2 + \cdots$$

that is converting it to values between 0 and 1. (probabilities). It is

$$\sigma(x) = \frac{1}{1 + e^{-x}}$$

Therefore we have

$$\frac{1}{1 + e^{-\sum_i x_i \omega_i}}$$

LUNAR CYCLE TRACKER

Fill in the phases of the moon as you observe them each night.

activation Function
(from wikipedia) —

"In computational networks,
the activation function of
a node defines the output of
that node given an input or
set of inputs. A standard
computer chip circuit can be
seen as a digital network of
activation Functions that can be
"ON" (1) or "OFF"(0), depending
on input."

Typical activation functions
of artificial neurons are
the sigmoid function

$$\sigma(x) = \frac{1}{1 + e^{-x}}$$

and, softmax

$$softmax(L_n) = \frac{e^{L_n}}{\| e^L \|}$$

the basic formula for
one layer of a neural network is:

$$Y = softmax(x \cdot \omega + b)$$

LUNAR CYCLE
TRACKER

Fill in the phases of the moon as you
observe them each night.

The basic method of creating
an artificial neuron is

to provide its input and
output for training sets,

do a sum over the weighted
values and feed that

through an activation function.

then compute the distances

between the actual and
computed values so you

can train the neuron or,
neural network. Once it is

trained it ready to interpret
new situations. The code in
Python for the sigmoid would be:

output = $1/(1+exp(-(dot(inputs, weights$

and using python's neural network
library:

output = $1/1 + np.exp(-model)$ where

model = $np.add(np.dot(x, weight))$
we assume bias is always zero

To compute the distance between what the system predicts and what we know to be true, we can use the cross entropy, then train the system as to what the weights and biases are:

$$\text{cross entropy} = -\sum Y_i' \cdot \log(Y_i)$$

Y_i' = actual probability

Y_i = computed probability

For the sigmoid function

$$\frac{d\,\sigma(x)}{dx} = \frac{1}{1+e^{-x}} = \left(\frac{1}{1+e^{-x}}\right)^2 \frac{d}{dx}(1+e^x)$$

$$= \left(\frac{1}{1+e^{-x}}\right)^2 e^{-x}(-1) = \left(\frac{1}{1+e^{-x}}\right)\left(\frac{1}{1+e^{-x}}e^{-x}\right)$$

$$= \left(\frac{1}{1+e^{-x}}\right)\left(\frac{-e^{-x}}{1+e^{-x}}\right) = \sigma(x)(1 - \sigma(x))$$

error = computed - actual

(error)(output)(1 - output)

math

$$input = \begin{matrix} a_{00} & a_{01} & a_{02} \\ a_{10} & a_{11} & a_{12} \\ a_{20} & a_{21} & a_{22} \end{matrix}$$

$$output = \begin{matrix} b_0 \\ b_1 \\ b_2 \end{matrix}$$ in code

input = array([[a_{00}, a_{01}, a_{02}], [a_{10}, a_{11}, a_{12}]...

output = array([[b_0, b_1, b_2]]).T

dot T (.T) transposes the
horizontal matrix into a vertical one

To perform the sum (math)

$$\sum_i x_i \omega_t = x_1 \omega_1 + x_2 \omega_2 + \ldots$$

100 times, write (code)

for iteration in xrange(100)

To perform the sum (code)

weights += dot(inputs.T, adjustment)

dot multiplies the two matrices
inputs.T and adjustment

Artificial Neural Networks

Milo Spencer-Harper, in his blog, learned to make a neural network from Andrew Trask, and I in turn learned from him in this example where he builds a neuron with python from scratch without using a neural network library, which makes this highly complex subject very accessible. He, in fact, reduced it to its simplest terms, and since we want the code in its simplest terms, for the next step of this project, we will use his code, but change the some of the variable names so we can see, more easily, what is going on, and so we can more easily talk about what we are doing.

To make a computer that thinks, we must make a computer that learns. To make a computer that learns, we have to give it experiences. This is called training the computer. To train it, we give it a set of problems, and provide the answers. To make a computer that thinks for itself, we have to let it make a random guess at the answers, then learn from its errors. This is done by the technique of creating a neural network that imitates the way the human mind works. In the human brains, synapses fire and, if they are strong enough, the neuron fires, and a thought is formed. To make a neural network, we must first make a single neuron. It is believed that if we can make a large enough network of neurons, that we can make something as complex as consciousness. It is my belief that the true problem of making something as self-contradictory as consciousness is to create something as paradoxical as a true random number generator. I say paradoxical because we don't have a true random number generator. Our current random number generators work according to a scheme and random, by definition, is that which has no scheme. I our generators are based on a method, it is just its output appears so unpredictable that we say, for all practical purposes, it is random. I believe the way our consciousness springs up is in that strange state of unpredictability that occurs in quantum mechanics. This is why I believe that quantum computers will be the way to tackle the problem. Not that neural networks won't play their role, but that the source of their input will be in that state of indeterminism.

The problem, as was put forward by Spencer-Harper and Track, and solved here by Spencer-Harper, will be four data sets are produced: (0, 0, 1), (1, 1, 1), (1, 0, 1), (0, 1, 0) with the respective answers (0, 1, 1, 0). We notice the respective answers are the left most numbers in the data sets. But, does the computer know this? Can it learn it, and produce the correct answer from a new data set outside the above experience? It can, and it is done as follows:

computer experiences	learns
0, 0, 1	0
1, 1, 1	1
1, 0, 1	0
0, 1, 0	0

new experiences	from what it learned it should deduce
1, 1, 0	1 or some number close to it (greater than 0.5, less than 1)
0, 1, 0	0 or some number close to it (less than 0.5, greater than 0)

We produce this (make an electronic neuron) as follows…

Use the random function with any seed (1 will do fine)

random.seed(1)

This says the computer does not know (it just guesses anything).

Then calculate the error in its guesses and iterate 10,000 times. Eventually the computer will learn a value close to the answer given in the training examples:

$$\sum_{1}^{10000} W_i I_i = W_1 I_1 + W_2 I_2 + W_3 I_3 + ...$$

The weighted input has to be between 0 and 1 so normalize the above with the sigmoid function S(x):

$$S(x) = \frac{1}{1+e^{-x}} \quad so \ that \ Output = \frac{1}{1+e^{-\sum_i W_i I_i}}$$

Calculate the error between the computer's result and that in the training example, and adjust the weights accordingly with the Error Weighted Derivative Formula...

Adjustment = (error)(output)($\nabla S(x)$) or, in other words,...

Adjustment = (learn−output)(output)(1−output)

The code looks like this in python:

```
from numpy import exp, array, random, dot
actual_in=array([[0,0,1],[1,1,1],[1,0,1],[0,1,1]])
actual_out=array([[0,1,1,0]]).T
random.seed(1)
weights=2*random.random((3,1))-1
for iteration in xrange(10000):
  computed=1/(1+exp(-(dot(actual_in, weights))))
  weights += dot(actual_in.T, (actual_out-computed)*computed*(1-computed))
  answer = (1/(1+exp(-(dot(array([1,1,0]), weights)))))
  print (answer)
if (answer<0.5):
    print ("The answer is 0")
else:
    print ("the answer is 1")
```

Running it in the Canopy shell.....

```
1 from numpy import exp, array, random, dot
2 actual_in=array([[0,0,1],[1,1,1],[1,0,1],[0,1,1]])
3 actual_out=array([[0,1,1,0]]).T
4 random.seed(1)
5 weights=2*random.random((3,1))-1
6 for iteration in xrange(10000):
7     computed=1/(1+exp(-(dot(actual_in, weights))))
8     weights += dot(actual_in.T, (actual_out-computed)*computed*(1-computed))
9     answer = (1/(1+exp(-(dot(array([1,1,0]), weights)))))
10    print (answer)
```

Editor - Canopy

artificial_neuron.py

Python   /Users/ianbeardsley

```
[ 0.99992234]
[ 0.99992235]
[ 0.99992236]
[ 0.99992236]
[ 0.99992237]
[ 0.99992238]
[ 0.99992239]
[ 0.9999224]
[ 0.9999224]
[ 0.99992241]
[ 0.99992242]
[ 0.99992243]
[ 0.99992244]
[ 0.99992244]
[ 0.99992245]
[ 0.99992246]
[ 0.99992247]
[ 0.99992248]
[ 0.99992248]
[ 0.99992249]
[ 0.9999225]
the answer is 1

In [2]:
```

Cursor pos   1 : 1   Python   ~/Desktop/artificial_neuron.py

Returning To My Theory

Visible spectrum $\approx 1\mu - 10\mu$
= near ultraviolet - near infrared

A germanium diode turns on at
0.3 Volts. There are $1.6E-19$ Joules/volt

$(0.3 V)(1.6E-19 \text{ joules/volt}) = 4.8E-20$ Joules.
We have said a silicon diode turns
on at 0.6 volts. So there are

$(0.6V)(1.6E-19 \text{ joules/volt}) = 9.6E-20$ Joules

we called that 1 bue (basic unit of energy)

Planck's Equation: $E = h\nu$, $h = 6.626E-34$

$\nu = 2(9.6E-20)/6.626E-34 = 3.067E14 \frac{\text{cycles}}{\text{sec}}$

$C = 3E8 \, {}^{m}/_{s}$ $\lambda = \dfrac{3E8}{3.067E-14} = 9.78E-7 \text{ meters}$

1 micrometer $= 1E-6$ meters

$\lambda \approx 1$ micrometer   This is where the visible
spectrum begins
near ultra violet 10μ

$\nu = 2(4.8E-20 \text{ Joules})/6.626E-34 \text{ J-sec}$

$= 1.4488E14 \frac{\text{cycles}}{\text{sec}} \approx 1.45E14 \frac{\text{cycles}}{\text{sec}}$

$\lambda = \dfrac{3E8}{1.45E14} = 2.1 \approx 2.0E-6 \text{ meters}$

$\lambda \approx 2.0$ micrometers   This is in the
visible spectrum

Modeling The Future

In 2009 I came up with the idea that we could determine a growth rate constant for the rate at which humanity develops. The idea was that since the stars system alpha centauri is about four light years away, is the closest star system to us, has a star the same spectral type as the sun and is the third brightest star in the sky while, the earth is the third planet from the sun that, if, we do a random walk to alpha centauri, that the percent probability of landing on alpha centauri in 10 random leaps of a light year each, was the percent development over a given period of time, of humanity. I just needed the timespan over which the development occurred. I knew that timespan began in 1969, because that was the year humans first set foot on the moon. I needed the end year and, I chose 2009 because, that is when we put telescopes into orbit that would see back to the beginning of the Universe. That 40 year period of development gave a growth rate constant of 0.0621. As it turned out, 0.0621 is the novelty rating in the McKenna Timewave for the year humans landed on the moon.

The percent development over that time period was 11.718750 as run in my C program on the C emulator. My program then rounded that value to the nearest integer, which was 12%. This is interesting, because I have since written a program on my hp 35s scientific calculator, and run the program for 12 hyperspatial jumps instead of 10. I originally chose ten because, our number system is base ten and we have ten fingers to count on. 12 hyperspatial jumps gives a chance of landing at plus four of 12%. It is the same value, but you don't have to round, and the growth rate constant is the same! I looked at 12 jumps because, 12 is the smallest abundant number, which is to say it is small but is evenly divisible by so many numbers 1, 2, , 3, 4, 6, 12. So many numbers that, 1+2+3+4+6=16 which is greater than 12 itself.

Today (January 05, 2017), I have revisited my C program, modelfuture.c and, honed it, by removing some curly brackets, which allow it to print out the results without printing out the calculations in-between to get those results. When running the program for 10 jumps you have to jump left 3, right 7, to land at 4, but, when running it for 12 jumps, you have to jump left 4, right 8, to land at 4. Here is the refined program in C and, running it for 12 jumps:

```c
#include <stdio.h>
#include <math.h>
int main (void)
{
printf("\n");
int N, r;
double u, v, y, z;
double t,loga, ratio;
int n1, n2;
char name[15];
float W,fact=1,fact2=1,fact3=1,a,g,rate,T,T1;
double x,W2;
printf("(p^n1)(q^n2)[W=N!/(n1!)(n2!)]");
printf("\n");
printf("x=e^(c*t)");
printf("\n");
printf("W is the probability of landing on the star in N jumps.\n");
printf("N=n1+n2, n1=number of one light year jumps left,\n");
printf("n2=number of one light year jumps right.\n");
printf("What is 1, the nearest whole number of light years to the
star, and\n");
printf("2, what is the star's name? (one word)\n");
printf("Enter 1: ");
scanf("%i", &r);
printf("Enter 2: ");
scanf("%s", name);
printf("Star name: %s\n", name);
printf("Distance: %i\n", r);
printf("What is n1? ");
scanf("%i", &n1);
printf("What is n2? ");
scanf("%i", &n2);
printf("Since N=n1+n2, N=%i\n", n1+n2);
N=n1+n2;
printf("What is the probability, p(u), of jumping to the left? ");
scanf("%lf", &u);
printf("What is the probability, p(v), of jumpint to the left? ");
scanf("%lf", &v);
printf("What is the probability, q(y), of jumping to the right? ");
scanf("%lf", &y);
printf("What is the probability, q(z), of jumping to the right? ");
scanf("%lf", &z);
printf("p=u:v");
printf("\n");
printf("q=y:z");
printf("\n");
```

```
for (int i=1; i<=N; i++)
{
fact = fact*i;
}
printf("N factorial = %f\n", fact);

a=pow(u/v,n1)*pow(y/z,n2);

for (int j=1; j<=n1; j++)
{
fact2 = fact2*j;
}
printf("n1 factorial = %f\n", fact2);

for (int k=1; k<=n2; k++)
{
fact3 = fact3*k;
}
printf("n2 factorial = %f\n", fact3);

x=2.718*2.718*2.718*2.718*2.718;
g=sqrt(x);
W=a*fact/(fact2*fact3);
printf("W=%f percent\n", W*100);
W2=100*W;
printf("W=%.2f percent rounded to nearest integral\n", round(W2));

{
printf("What is t in years, the time over which the growth occurs? ");
scanf("%lf", &t);
loga=log10(round(W*100));
printf("log(W)=%lf\n", loga);
ratio=loga/t;
printf("loga/t=%lf\n", ratio);
rate=ratio/0.4342; //0.4342 = log e//
printf("growthrate constant=%lf\n", rate);
printf("log 100 = 2, log e = 0.4342, therfore\n");
printf("T=2/[(0.4342)(growthrate)]\n");
T=2/((0.4342)*(rate));
printf("T=%.2f years\n", T);
printf("What was the begin year for the period of growth? ");
scanf("%f", &T1);
printf("Object achieved in %.2f\n", T+T1);
}
}
```

Here we run it in Harvard's jharvard appliance:

jharvard@appliance (~): cd Dropbox
jharvard@appliance (~/Dropbox): make modelfuture
clang -ggdb3 -O0 -std=c99 -Wall -Werror   modelfuture.c -lcs50 -lm -o modelfuture
jharvard@appliance (~/Dropbox): ./modelfuture

$(p^{n1})(q^{n2})[W=N!/(n1!)(n2!)]$
$x=e^{(c*t)}$
W is the probability of landing on the star in N jumps.
N=n1+n2, n1=number of one light year jumps left,
n2=number of one light year jumps right.
What is 1, the nearest whole number of light years to the star, and
2, what is the star's name? (one word)
Enter 1: 4
Enter 2: alphacentauri
Star name: alphacentauri
Distance: 4
What is n1? 4
What is n2? 8
Since N=n1+n2, N=12
What is the probability, p(u), of jumping to the left? 1
What is the probability, p(v), of jumpint to the left? 2
What is the probability, q(y), of jumping to the right? 1
What is the probability, q(z), of jumping to the right? 2
p=u:v
q=y:z
N factorial = 479001600.000000
n1 factorial = 24.000000
n2 factorial = 40320.000000
W=12.084961 percent
W=12.00 percent rounded to nearest integral
What is t in years, the time over which the growth occurs? 40
log(W)=1.079181
loga/t=0.026980
growthrate constant=0.062136
log 100 = 2, log e = 0.4342, therfore
T=2/[(0.4342)(growthrate)]
T=74.13 years
What was the begin year for the period of growth? 1969
Object achieved in 2043.13
jharvard@appliance (~/Dropbox):

Here I run the program for 12 jumps in the Mac Utility Terminal, after compiling it with X-Code Beta:

```
Last login: Thu Jan  5 17:28:51 on ttys000
Claires-MBP:~ ianbeardsley$ /Users/ianbeardsley/Desktop/modelfuture ;
exit;

(p^n1)(q^n2)[W=N!/(n1!)(n2!)]
x=e^(c*t)
W is the probability of landing on the star in N jumps.
N=n1+n2, n1=number of one light year jumps left,
n2=number of one light year jumps right.
What is 1, the nearest whole number of light years to the star, and
2, what is the star's name?
Enter 1: 4
Enter 2: alphacentauri
Star name: alphacentauri
Distance: 4
What is n1? 4
What is n2? 8
Since N=n1+n2, N=12
What is the probability, p(u), of jumping to the left? 1
What is the probability, p(v), of jumpint to the left? 2
What is the probability, q(y), of jumping to the right? 1
What is the probability, q(z), of jumping to the right? 2
p=u:v
q=y:z
N factorial = 479001600.000000
n1 factorial = 24.000000
n2 factorial = 40320.000000
W=12.084961 percent
W=12.00 percent rounded to nearest integral
What is t in years, the time over which the growth occurs? 40
log(W)=1.079181
loga/t=0.026980
growthrate constant=0.062136
log 100 = 2, log e = 0.4342, therfore
T=2/[(0.4342)(growthrate)]
T=74.13 years
What was the begin year for the period of growth? 1969
Object achieved in 2043.13
2017-01-05 17:30:46.494 modelfuture[11820:23612253] Hello, World!
logout

[Process completed]
```

My Earlier Work in Model Future

Model Future Short

We equate the probability of landing at a distance in a random walk to the percent development of Humanity. Thus, as an example, we say landing at plus 4 after 10 random jumps of one light year each with equal probability of jumping either left or right (p=0.5, q=0.5) which can be expressed (p =1/2, q=1/2) that there is a 12% development towards building a starship since 1969 (The year we landed on the moon) to 2009 (the year we put space telescopes in orbit that would see back to the beginning of the Universe). 2009 - 1969 = 40 years. Thus in 40 years we have developed 12% towards the objective. Knowing the t = 40 years we can use the equation of exponential growth to determine the growth rate constant, k. Which in this case is 0.0621. Knowing the growth rate, we can determine in what year we will have a starship since 1969 is time = 0. We can adjust the probabilities towards the objective and away from the objective according to our assessment of the human state at any given moment in time. The following is the equation of a random walk. N is the number of jumps (10), p and q are the probabilities of jumping either left or right, and n1 and n2 are the jumps left and right. So if we are to land at plus 4 in 10 leaps, then n1 = 3 and n2 =7 (n1+n2 = 10, and n2-n1=7).

$$W_n(n1) = \frac{N!}{n_1! n_2!} p^{n1} q^{n2}$$

$$k = \frac{\log W_n(n_1)}{T(\log 2.718)}$$

$$T = \frac{\log 100}{k(\log 2.718)}$$

W:HP   Part 1   Percent Development   hp 35s

W001 LBL W
W002 INPUT N
W003 (
W004 STO X
W005 INPUT A
W006)
W007 STO Y
W008 INPUT B
W009)
W010 STO Z
W011 INPUT P
W012 RCL A
W013 y^x
W014 STO U
W015 INPUT Q
W016 RCL B
W017 y^x
W018 STO V
W019 RCL U
W020 RCL V
W021 X (MULTIPLY)
W022 STO C
W023 RCL Y
W024 RCL Z
W025 X (MULTIPLY)
W026 STO D
W027 RCL X
W028 RCL C
W029 X (MULTIPLY)
W030 RCL D
W031 / (DIVIDE)
W032 100
W033 X (MULTIPLY)
W034 RTN

To run this, type:

XEQ W000
N? 10   R/S
A? 3   R/S
B? 7   R/S
P? 0.5   R/S
Q? 0.5   R/S

      RUNNING

Answer: 11.2332   Which is correct, it rounds to 12%

The HP 35s

W.HP   Part 1   Percent Development   hp 35s

W001 LBL W
W002 INPUT N
W003 !
W004 STO X
W005 INPUT A
W006 !
W007 STO Y
W008 INPUT B
W009 !
W010 STO Z
W011 INPUT P
W012 RCL A
W013 y^x
W014 STO U
W015 INPUT Q
W016 RCL B
W017 y^x
W018 STO V
W019 RCL U
W020 RCL V
W021 X (MULTIPLY)
W022 STO C
W023 RCL Y
W024 RCL Z
W025 X (MULTIPLY)
W026 STO D
W027 RCL X
W028 RCL C
W029 X (MULTIPLY)
W030 RCL D
W031 / (DIVIDE)
W032 100
W033 X (MULTIPLY)
W034 RTN

To run this, type:

XEQ W000
N? 10   R/S
A? 3   R/S
B? 7   R/S
P? 0.5   R/S
Q?i 0.5   R/S

...... RUNNING

Answer: 11 23/32   Which is correct, it rounds to 12%

R.HP   Part 2   Growth Rate   hp 35s (Using result from part 1, W.HP)

R001 LBL R
R002 INPUT W
R003 LOG
R004 STO W
R005 INPUT T
R006 STO T
R007 2.718
R008 LOG
R009 STO L
R010 RCL W
R011 RCL T
R012 / (DIVIDE)
R013 RCL L
R014 / (DIVIDE)
R015 RTN

To run this, type:

XEQ R000   R/S
W? 12   R/S
T? 40   R/S

…… RUNNING

Answer: 178/2865 = 0.0621 (Which is Correct)

T.HP   Part 3   Time Objective Achieved hp 35s (Using results from part 2, R.HP)

T001 LBL T
T002 INPUT R
T003 STO R
T004 100
T005 LOG
T006 STO L
T007 2.718
T008 LOG
T009 STO E
T010 RCL L
T011 RCL R
T012 / (DIVIDE)
T013 RCL E
T014 / (DIVIDE)
T015 RTN

To run, type:

XEQ T000   R/S
R? 0.0621   R/S

...... RUNNING

Answer: 74 167/1012 =74.165 years ~ 74 years (Which is correct)

1969 + 74 =2043

Star System: Alpha Centauri
Spectral Class: Same As The Sun
Proximity: Nearest Star System
Value For Projecting Human Trajectory: Ideal

The probability of landing at four light years from earth at Alpha Centauri in 10 random leaps of one light year each (to left or right) is given by the equation of a random walk:

{ W }_{ n }({ n }_{ 1 })=\frac { N! }{ { n }_{ 1 }!{ n }_{ 2 }! } { p }^{ n1 }{ q }^{ n2 }\\
N={ n }_{ 1 }+{ n }_{ 2 }\\ q+p=1

$$W_n(n_1) = \frac{N!}{n_1! n_2!} p^{n1} q^{n2}$$

$$N = n_1 + n_2$$

$$q + p = 1$$

To land at plus four we must jump 3 to the left, 7 to the right (n1=3, n2 = 7: 7+3=10):

Using our equation:

$$\frac{(10!)}{(7!)(3!)}(\frac{1}{2})^7(\frac{1}{2})^3 = \frac{3628800}{(5040)(6)}\frac{1}{128}\frac{1}{8} = \frac{120}{1024} = \frac{15}{128} = 0.1171875 \approx 12\%$$

We would be, by this reasoning 12% along in the development towards hyperdrive.

Having calculated that we are 12% along in developing the hyperdrive, we can use the equation for natural growth to estimate when we will have hyperdrive. It is of the form:

$$x(t) = x_0 e^{kt}$$

t is time and k is a growth rate constant which we must determine to solve the equation. In 1969 Neil Armstrong became the first man to walk on the moon. In 2009 the European Space Agency launched the Herschel and Planck telescopes that will see back to near the beginning of the universe. 2009-1969 is 40 years. This allows us to write:

$$12\% = e^{k(40)}$$

$$\log 12 = 40k \log 2.718$$

$$0.026979531 = 0.4342 k$$

$$k=0.0621$$

We now can write:

$$x(t) = e^{(0.0621)t}$$

$$100\% = e^{(0.0621)t}$$

$$\log 100 = (0.0621) t \log e$$

$$t = 74 \text{ years}$$

$$1969 + 74 \text{ years} = 2043$$

Our reasoning would indicate that we will have hyperdrive in the year 2043.

Study summary:

1. We have a 70% chance of developing hyperdrive without destroying ourselves first.
2. We are 12% along the way in development of hyperdrive.
3. We will have hyperdrive in the year 2043, plus or minus.

Sierra Waters was handed the newly discovered document in 2042. Sierra Waters was a Paul Levinson character in his book The Plot To Save Socrates. Her adventure was one to excel human development, in particular to attain hyperdrive technology for making a starship.

In Isaac Asimov's I, Robot, he has that a computer called The Brain will invent hyperdrive in 2044. My value of 2043 is right in-between these two assessments.

The Perceptron

Instead of building our neuron for a specified ~~training set~~ training set, we can build it for any training set, so we have a summing node that ~~applies~~ computes a linear combination of the inputs applied to the synapses with a bias. We apply this sum to a hard limiter. The neuron produces an output of $+1$ if the hard limiter input is is positive and -1 if it is ~~negative~~. We denote the weights as $w_1, w_2, \cdots w_i$ and the inputs as $x_1, x_2, \cdots x_i$. This is Rosenblatt's Perceptron. Rosenblatt (1958, 1962) developed his perceptron brain model as a learning procedure for patterns (vectors). It is simply,

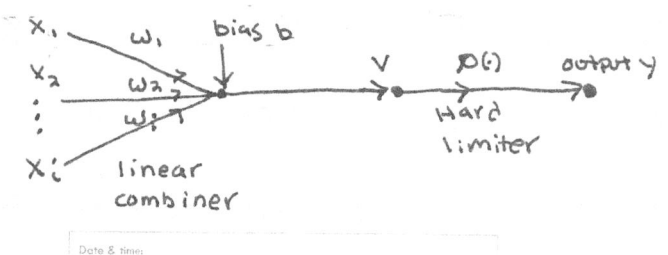

$$v = \sum_{i=1}^{r} \sum w_i x_i + b$$

the bias, b, is a constant particular to the system. Since $w_i x_i$ is the product of two matrices often of more than 1 dimension, then how can you add b of different dimensions?

It is often dealt with by adding it ~~\|\|\|\|~~ to each of the elements:

$$v = \sum_{i=1}^{n} (w_i x_i + b) \qquad v(n) = \sum_{i=0}^{m} \omega_i(n) x_i(n) = w^T(n) x(n)$$

If weight vector member $x(n)$ is correctly classified by the weight vector $w(n)$ no correction is made:

~~w(n+1) = w(n) if w(n)x(n)~~

$w(n+1) = w(n)$ if $w^T(n) x(n) > 0$ $x(n)$ of class 1
$w(n+1) = w(n)$ if $w^T(n) x(n) \leq 0$ $x(n)$ of class 2

otherwise it is updated as follows

$w(n+1) = w(n) - \eta(n) x(n)$ if $w^T(n) x(n) > 0$ $x(n)$ of class 2
$w(n+1) = w(n) - \eta(n) x(n)$ if $w^T(n) x(n) \leq 0$ $x(n)$ of class 1

$\eta(n)$ is the learning rate parameter that controls ~~the weight vector at~~ the adjustment to the weight vector at the ith iteration.

we can draw the schematic for
Rosenblatt's perceptron in terms of
what we already know:

inputs   weights

x_1 w_1 net   activation
                  input   Function
                  function
x_2 w_3 $\sum \longrightarrow \sigma(x) \longrightarrow$ output
$\vdots$ w_i
x_i

Date & time:

Date & time:

Location

Orientation

I have built a perceptron that produces three random number between 0 and 1 and that then asks you for three values: input, weight, and bias. You can use the random values if desired.

```python
import numpy as np
from numpy import random, dot
print ("three random probabilities: ")
print 2*random.random(3)-1;
x=float(raw_input("Give me your input: "))
weight=float(raw_input("Give me the weight: "))
bias=float(raw_input("Give me the bias: "))
model=np.add(np.dot(x, weight), bias)
activation=1/(1+np.exp(-model))
print("weighted inputs plus bias passed through sigmoid yield a probability: ")
print activation
```

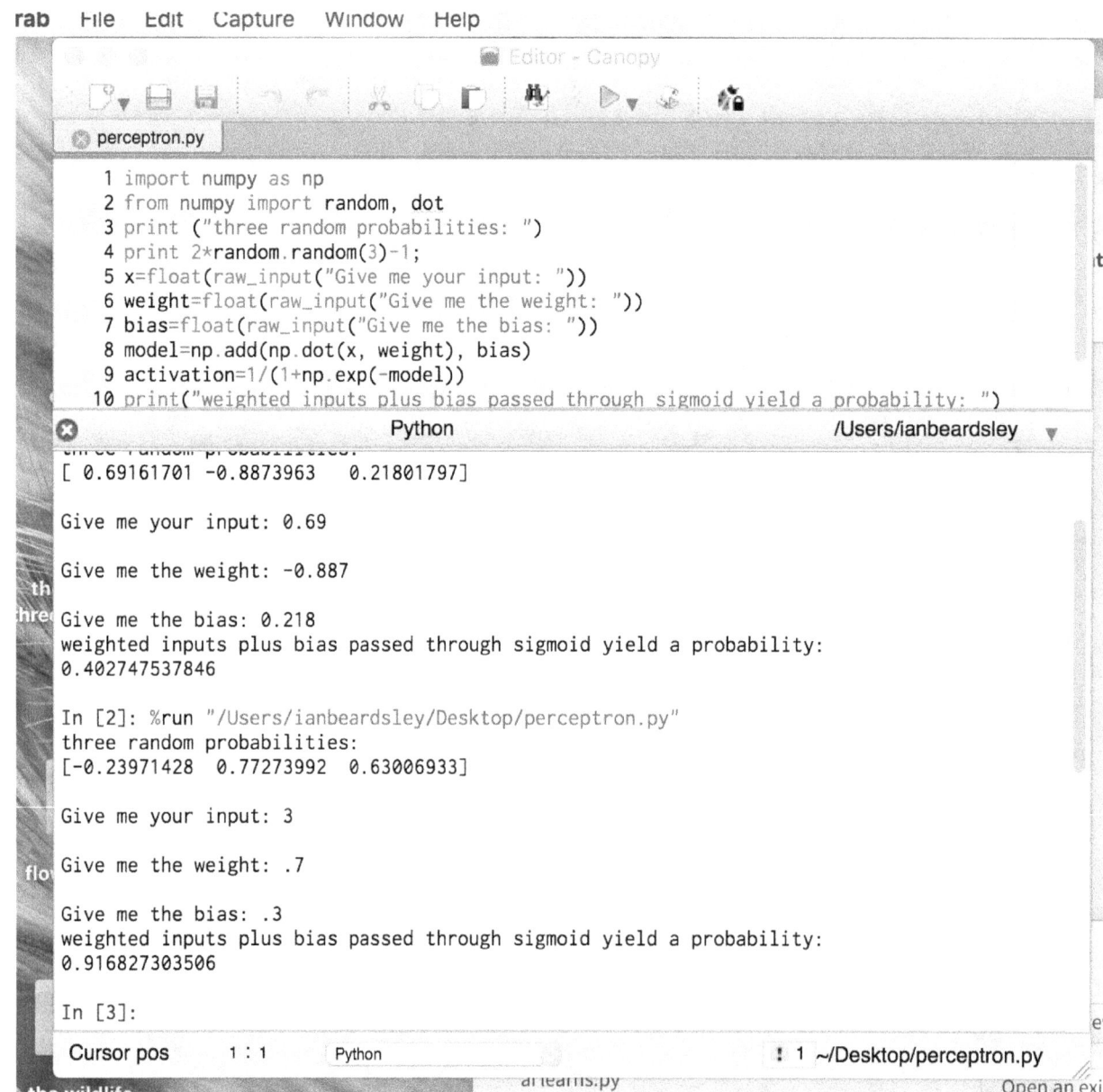

```
1 import numpy as np
2 from numpy import random, dot
3 print ("three random probabilities: ")
4 print 2*random.random(3)-1;
5 x=float(raw_input("Give me your input: "))
6 weight=float(raw_input("Give me the weight: "))
7 bias=float(raw_input("Give me the bias: "))
8 model=np.add(np.dot(x, weight), bias)
9 activation=1/(1+np.exp(-model))
10 print("weighted inputs plus bias passed through sigmoid yield a probability: ")
```

```
[ 0.69161701 -0.8873963   0.21801797]

Give me your input: 0.69

Give me the weight: -0.887

Give me the bias: 0.218
weighted inputs plus bias passed through sigmoid yield a probability:
0.402747537846

In [2]: %run "/Users/ianbeardsley/Desktop/perceptron.py"
three random probabilities:
[-0.23971428  0.77273992  0.63006933]

Give me your input: 3

Give me the weight: .7

Give me the bias: .3
weighted inputs plus bias passed through sigmoid yield a probability:
0.916827303506

In [3]:
```

I run it for my growth rate constant 0.0621 and the 12% growth.

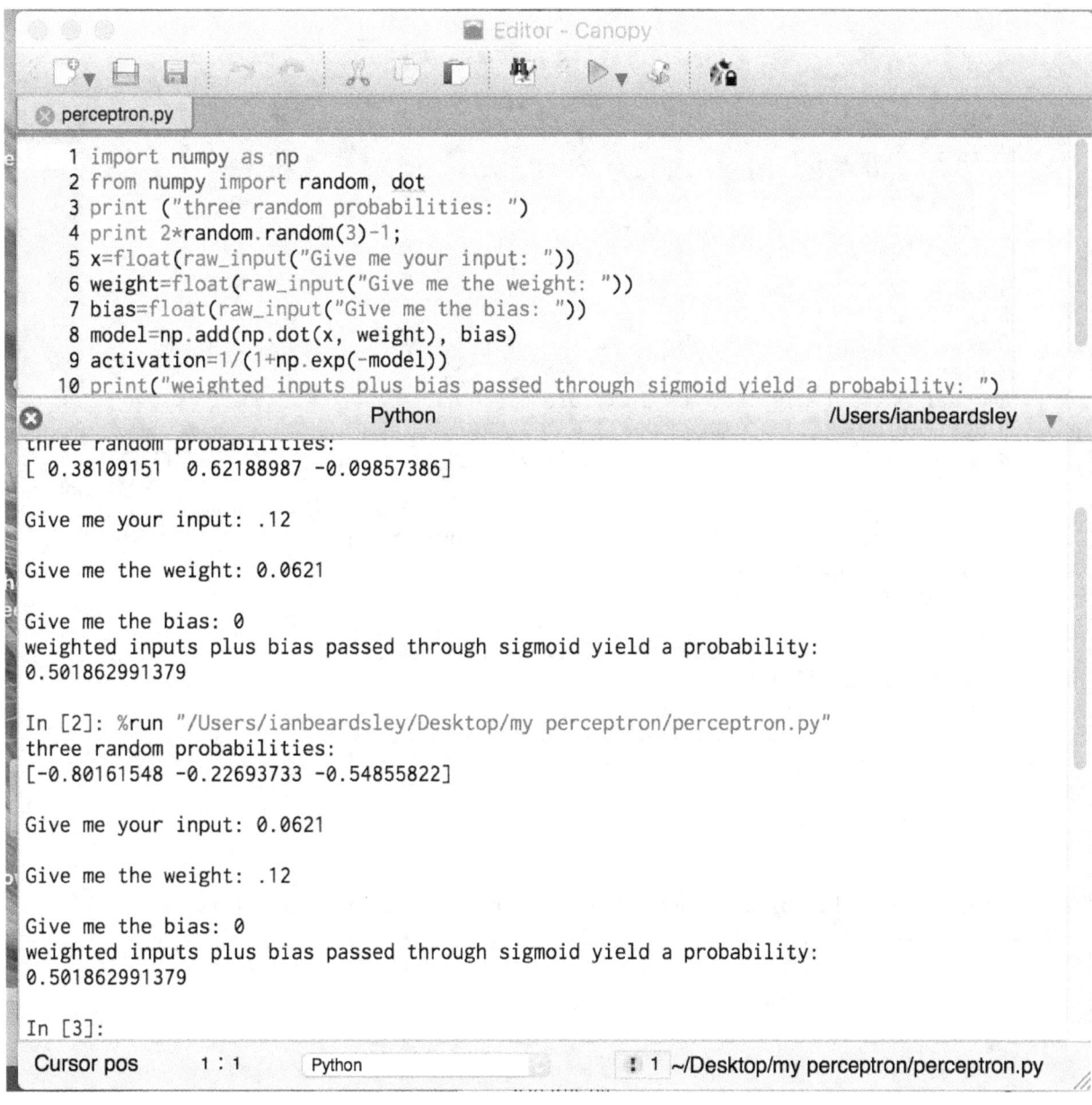

```
1 import numpy as np
2 from numpy import random, dot
3 print ("three random probabilities: ")
4 print 2*random.random(3)-1;
5 x=float(raw_input("Give me your input: "))
6 weight=float(raw_input("Give me the weight: "))
7 bias=float(raw_input("Give me the bias: "))
8 model=np.add(np.dot(x, weight), bias)
9 activation=1/(1+np.exp(-model))
10 print("weighted inputs plus bias passed through sigmoid yield a probability: ")
```

```
three random probabilities:
[ 0.38109151  0.62188987 -0.09857386]

Give me your input: .12

Give me the weight: 0.0621

Give me the bias: 0
weighted inputs plus bias passed through sigmoid yield a probability:
0.501862991379

In [2]: %run "/Users/ianbeardsley/Desktop/my perceptron/perceptron.py"
three random probabilities:
[-0.80161548 -0.22693733 -0.54855822]

Give me your input: 0.0621

Give me the weight: .12

Give me the bias: 0
weighted inputs plus bias passed through sigmoid yield a probability:
0.501862991379

In [3]:
```

Running at 12%

perceptron.py

```
1 import numpy as np
2 from numpy import random, dot
3 print ("three random probabilities: ")
4 print 2*random.random(3)-1;
5 x=float(raw_input("Give me your input: "))
6 weight=float(raw_input("Give me the weight: "))
7 bias=float(raw_input("Give me the bias: "))
8 model=np.add(np.dot(x, weight), bias)
9 activation=1/(1+np.exp(-model))
10 print("weighted inputs plus bias passed through sigmoid yield a probability:
```

Python   /Users/ianbeard

```
Welcome to Canopy's interactive data-analysis environment!
 with pylab-backend set to: qt
Type '?' for more information.

In [1]: %run "/Users/ianbeardsley/Desktop/my perceptron/perceptron.py"
three random probabilities:
[-0.57840598 -0.61031973 -0.06469468]

Give me your input: 12

Give me the weight: 0.0621

Give me the bias: 0
weighted inputs plus bias passed through sigmoid yield a probability:
0.678131904912

In [2]:
```

57

A440 Connected

The growth rate constant, $k=0.0621$ has turned up in other places in my work. In my work *Theory of Structure in AI* I talked about 9/5 and how 9 and 5 are the yin and yang. We write:

$$\frac{9}{5} = yin \qquad \frac{5}{3} = yang \qquad 621 = L \text{ (Levinson's Number)}$$

$15 = M$ (Manuel's Number)

$$1080 \ yang + \frac{35}{3} \ yin = L$$

$$100 \ yin \frac{120 \ yang + 7}{4} = ML$$

120 is the angle of the yang as derived from 5/3 and is the 120 degrees in a regular hexagon, which is six fold symmetry.

$$100\left(\frac{9}{5}\right) \frac{\left(\frac{5}{3}\right)\left(\frac{5}{3}\right) + 7}{5} = 15L$$

$$L = 29\frac{1}{3}$$

$$M = 621/440 \approx \sqrt{2}$$

Let us invert that and redifine M as $\dfrac{\sqrt{2}}{2}$

This is an important angle $\sin 45 = \cos 45 = \dfrac{\sqrt{2}}{2}$
45 degrees is the angle for maximum range of a projectile.

And, $440 = ML$

We notice $(23)(27) = 440$

3 cubed is 27.

The Earth rotates through $\dfrac{360}{24}=15$ degrees/hour

$$\dfrac{15\ deg}{hour}\ \dfrac{hour}{60\ min}\ \dfrac{min}{60\ sec}=0.0004166667\ degrees/second$$

That is $\dfrac{240\ seconds}{one\ degree}$

$$\dfrac{440\ cycles}{second}\ \dfrac{one\ degree}{440}=\dfrac{11}{6}$$

$$\dfrac{360}{6}=six\text{-}fold\ symmetry,\ 360-60=300\ =\ \dfrac{5}{6}\ \ \dfrac{5}{6}+1=\dfrac{11}{6}$$

11/6 is my other yang in other work as it is an alternate six-fold symmetry.

The Sothic Cycle is a cycle of the Egyptian Calendar, which is the time between heliacal rising of the brightes star, Sirius, which coincides with the annual inundation of the Nile river. It is 1,460 Julian years (365.25 days)

$$\frac{11}{6} = \frac{360 - \frac{360}{6}}{360} + 1$$

$$L = 621; \quad M = 0.708534622 \approx \frac{\sqrt{2}}{2}$$

$$L \approx 1000\phi = (1000)(goleden \ ratio \ conjugate)$$

$$\frac{A440}{240} = \frac{ML}{240} = \frac{11}{6}$$

$$\frac{24 \ hr}{day} \frac{60 \ min}{hr} \frac{1 \ year}{360 \ degrees} = \frac{1440}{day} \frac{year}{360} = \frac{4 \ min}{day}$$

$$\frac{240 \ sec}{deg} \frac{1 \ min}{60 \ sec} = \frac{4 \ min}{day}$$

$$\phi = 0.618$$

$$\frac{4 \ min}{day} \frac{365 \ day}{year} = 1,460 = sothic \ cycle$$

The Author

www.ingramcontent.com/pod-product-compliance
Lightning Source LLC
Chambersburg PA
CBHW081224170526
45165CB00009B/2936